FENCING MANUAL

Fencing Manual

Approved by the Minister of War,
18 May 1877

Manuel d'Escrime

Ministère de la Guerre

translated by Chris Slee

Fencing Manual
Copyright ©2017 Chris Slee (translator)

ISBN: 978-0-9943590-7-0 (eBook)
ISBN: 978-0-9943590-6-3 (Print)

Manuel d'Escrime, Ministère de la Guerre. The original text is from a facsimile of the 1877 edition. It is asserted this book is in the public domain.

Created at LongEdge Press, first edition

To all pioneers of Historical European
Martial Arts past and present

Contents

Introduction

The *Fencing Manual* 1877 stands as the gateway between "ancient" and modern fencing. The technical fencing terms used in the *Manual* are the same as those used by fencers today but the emphasis is on a set of core techniques more suitable to a weapon much heavier than the modern sport *épée* or sabre. It can be seen as the last of the manuals in France covering swordsmanship as a solely military activity concerned with life and death on the battlefield (as well as the dictates of nobility). But it can also be seen as the first of the modern fencing manuals in which swordsmanship has a sporting and even a civic purpose.

After defeat in the Franco-Prussian War of 1870, France was in upheaval. Before the war, in 1869, Emperor Napoleon III brought back to the military obligatory instruction in swordsmanship and this impulse survived and continued after the disastrous year following which saw not only France's defeat by the Prussian army but the abdication of the Emperor and founding of the Third Republic.

There was a strong popular movement to revive French national pride and fencing was seen as a key tool to achieve this. Some, such as General Georges Boulanger, sought to rebuild the French army in order for the country to revenge itself on the Prussian enemy. The Constitution of the Third Republic was officially enacted in 1875, a full five years after the Republic was declared. That the *Manual* was published within two years of this event shows, I think, it's importance to the national project.

The *Manual* was released under the authority of the Minister of War, Jean Auguste Berthaut, himself a commander of the National Guard in Paris in 1870. It was in use by the French army from the time of its publication in May 1877 for thirty years, with the occasional memos or official circulars clarifying ambiguities and providing direction, until the updated edition was published in 1908. It can be found listed in all the circulars outlining the books that officer candidates were expected to own and have read during this time.

Also during this period, the world of fencing changed radically. In France in the 1880s, duelling became once again a popular activity undertaken to redress grievances – particularly between politicians and the journalists who wrote about them.

This decade saw the flourishing of *salles d'armes* throughout the country as fencing became something of an *oeuvre patriotique* (patriotic endeavour) and ingrained into the educational program to improve the physical, intellectual and moral standing of French youth. Even the Joinville academy had by this time modified its original purpose of training army instructors and had opened its doors in a limited way to the public. In 1882, the *Société d'Encouragement à l'Escrime* (Society for the Encouragement of Fencing) was founded along with many other associations and umbrella bodies to codify how fencing schools were run. In 1906, the *Société* merged with the *Association Escrime Française* (Association of French Fencing) to form what we know as the modern FFE, the *Federation Français d'Escrime*.

The 1890s saw the first talk of an *escrime sportive* or sports fencing. In 1895, the magazine *Escrime Français* ran a hugely publicised tournament between top Italian and French fencers. In 1896, men's fencing was accepted as a demonstration sport in the first Olympic Games. By 1900, fencing had largely found the shape it has today.

The *Manual* itself is divided into two disciplines. The first part concerns itself with *épée* or fencing with the point – the weapon we usually associate with point-fencing, the foil, is not included. All army personnel, whether infantry or cavalry, were required to learn and practice with the *épée*. The second part is about *contre-point* or fencing with the edge for which the chief weapon was the military sabre. This was restricted to members of the cavalry arm.

The 1908 updated manual abandons this distinction in favour of officers of all arms training with the three modern weapons in use today, foil, *épée* and sabre.

In both parts, the learning progression begins with naming the parts of the weapon, stance and simple actions then through a series of set piece exercises of increasing complexity and difficulty until the student is judged ready for free sparring. One interesting feature here is the advice to instructors to find a way to adapt the training program to the needs of the individual student rather than to forcing all students into the same, one-size-fits-all mold.

Notes on the Text

The decision on whether or not to translate technical fencing terms is bound to draw criticism. I chose to translate many terms into English because I felt that the French term, although used today, either refers to an action which was significantly different to its modern definition or is performed in a different manner due to difference in the weight and performance of a military weapon over a modern sporting weapon. I do not claim to be an expert in either but the decision makes sense in my mind.

The layout of the exercises is terse and takes some experimentation to understand exactly the exercises are to be performed. Here's an example.

Parries and Ripostes of the Straight Strike

On a disengagement:

1. Engage the sword
2. *Sixte*, oppose (or parry) and riposte straight
3. *En garde*

After an opposition, parry with an opposition

2. Counter of *quarte*, oppose (or parry) and riposte straight
2. *Quarte*, oppose (or parry) and riposte straight
2. Counter of *sixte*, oppose (or parry) and riposte straight

After a counter, parry with a counter

The first itemised list is clear. It sets out three steps in order to achieve the exercise. This is the basic pattern of the exercises. Then comes a number of items labelled 2. These are options which are substituted in turn as the second action of the exercises in question. This is short example of the text's brevity. Some examples can continue for several pages with many intervening headings to pinpoint the skill or technique being demonstrated.

FENCING MANUAL

Basis of Instruction

Instruction in Fencing is divided into two parts each comprising six articles.

First Part

Fencing with the *épée* or point fencing

First Article: Outline of the method of teaching. The spirit in which it must be practiced.

Article 2: Definition of terms used in fencing and description of the positions which they explain and the movements which attach to them.

Article 3: Teaching progression.

Article 4: General rules to observe.

Article 5: Details of the lessons.

Article 6: The wall and the assault.

Second Part

Fencing with the sabre or *contre-point* fencing

First Article: Outline of the method of teaching. The spirit in which is must be practiced.

Article 2: Description of the sabre and of the positions and movements which attach to it to put it into play for attack and defence.

Article 3: Teaching progression.

Article 4: General rules to observe.

Article 5: Details of the lessons.

Article 6: The wall and the assault.

General Rules

Instruction in fencing is obligatory and free and ruled by the daily service table by the company, squadron or battery.

In each barracks, a special *salle d'armes* is assigned[1] to officers who must book regular and daily exercise in order to give the troop an example of the taste of fencing (Ministerial Circular of 7 May 1875). To this effect, the master of arms and the provosts shall be placed at their disposal at times fixed by the commander of the corps or detachment who must seek to put the weapons in the spotlight,[2] encouraging public bouts[3] and engaging the officers to attend and even participate in order to give them more solemnity.

In each company, squadron or battery, detached or forming the corps, the oversight and direction of the instruction is exercised by the commanding captain.

In each regiment, direction is exercised by a captain and general oversight by a superior officer.

The training personnel are constituted conforming to the guidelines[4] in the tables annexed to the laws of 13 March and of 15 December 1875 on army officers,[5] and the Ministerial Circular of 17 June 1876 on the organisation of the fencing service in the training squads of military crews.[6]

Troops on Foot

In foot troops, one teaches especially fencing with the *épée*.[7] Fencing with the sabre is only secondary and optional.

The instruction for the soldiers begins from their admission to the company's school and it is persued during the entire time of their presence under the flag.[8]

The timetable[9] fixes the hour each day assigned to individual companies who must follow each other in the *salle d'armes*, the manner that each company can go there, at least three times per week, without impeding other exercises. The men are conducted to the *salle d'armes*

---

[1] *affectué*
[2] *à mettre les armes en honneur*
[3] *les assauts publics*
[4] *indications*
[5] *sur les cadres de l'armée*
[6] *sur l'organisation de l'escrime dans les escadrons du train des équipages militaires*
[7] ie: a straight bladed weapon as opposed to the curved sabre
[8] ie: for the duration of their military service
[9] *le tableau de l'emploi du temps*

by the sargeant of the week, who notes the absent and reports to the officer of the week a scrupulous account of the reasons for absence.

The head of the corps also regulates the hours of the lessons of the sub-officiers who are authorised to procede individually to the *salle d'armes* and who must, like the troops, take at least three lessons per week.

If, *by exception*, insufficiency of available time or teaching personnel does not permit conforming strictly to the prescriptions above, one can bear *momentarily* to have the *salle* frequented only by the sub-officiers, corporals and student corporals[10] who must always be obliged[11] to undertake the lesson.

One carries on the men's notebooks their degree of fencing instruction.

One mentions, in the notes given to officers, their aptitude for fencing.

Mounted Troops

In the body of mounted troops, one gives concurrently instruction in fencing with the point and in fencing with the edge.[12]

Instruction for the soldiers begins from their admission to the squadron's school and it is persued during their entire presence under the flag.

The head of the corps must comply with the regulation of training with the dispositions prescribed above for troops on foot, appropriating the constitent elements of them to their troops.

---

[10] *les élèves caporaux*

[11] *être astreints*

[12] *la contre-pointe*

First Part: Fencing with the *Épée*

Article the First — Outline of the teaching method. The spirit in which it must be practiced.

Instruction in fencing with the sword is always individual.

It has as its aim teaching the student to direct his sword in attack and defence. It procedes from the simple to the complex[13] making understood by the student: a summary of sword nomenclature,[14] the manner of holding the sword, movements preparatory to taking guard, taking guard, gathering forward and backwards, stepping forward and backwards, the *appels*, postioning of the arms, the feint, the resumption of the guard, various means of attack and defence in order of the lessons, the progression, the wall, the salut and, finally, the assault.[15]

The instructor must adapt[16] his manner of acting to the temperament, character, disposition and intelligence of the student for whom he must identify the method to follow step-by-step [for] his development and to be used to achieve his strength, in all correcting without cease the defective positions of the hand and body before progressively accelerating the rapidity of the movements in order to ready[17] precision and speed, essential qualities of the fencer who must keep

---

[13] *composé*
[14] *la nomenclature sommaire de l'épée*
[15] ie: bouting, sparring and/or fighting in earnest
[16] *approprier*
[17] *préparer*

all his inidividuality and be be gradually led to implement,[18] with his maximum power, all the elements of action appropriate to him.

The instructor must have care to insist on the practical demonstration of first principles, requiring that they are strictly observed, that the guard above all is always correct and only pass on when the posture and natural movements of the student no longer call for any correction.

Each detail taught must find its application in the assault. The instructor will attempt therefore to highlight[19] its utility through practical demonstration, engaging[20] and repeated, accompanied by brief explanations, clear, clean and precise, allowing the student to realise[21] the correlation between attacks and defences and succeed, later, in making an intelligent and reasoned bout[22] where he will rule his game following the demands of those (the rules) of his opponent.

As and when the needs of instruction [dictate],[23] he will make known to the student the meaning of the usual terms of fencing. He will bring [him] thus to define and describe: the essential parts of the sword, the guard, the assembly, the lunge,[24] the different lines, the engagement, the changing of engagement, the double engagement, the fingering,[25] the attack and its variations, the right strike, disengagement, the cutover,[26] feint, the glide,[27] the beat, pressing, explusions,[28] the bind,[29] disengagement in time,[30] the *remise*, the *reprise*, the redouble, a timed strike, a stop-hit, the parry and its variations, opposition, the counter, riposte, counter-riposte, the fencing phrase.[31]

---

[18] *à mettre en oeuvre*
[19] *à en faire ressortir*
[20] *saisissante*
[21] *se rendre compte*
[22] *assaut*
[23] *au fur et à mesure des boins de l'instruction*
[24] *fente*
[25] *doigté*
[26] *coupé*
[27] *coulé*
[28] *froissement*
[29] *liement*
[30] *dérobement*
[31] *la phrase d'armes*

Art. II — Definition of the terms used in fencing and description of the postions which they explain and of the movements attached to them.

Épée

The sword, the intrument of fencing, is composed of two principle parts: the *blade* and the *mounting*.[32]

The blade, in steel, presents:

- the Point, the forward weak part terminated by the *button* or the *fly*[33] with which one hits;
- the Heel, the rearward strong part touching the mounting and with which one parries;
- the Middle, the intermediary middle part, immediately before which one must make the engagements.

The blade is attached to the mounting by the *tang*[34] which is rivetted on the *pommel* serving to counter-weight the blade.

The mounting comprises the *handle*[35] and the *guard*.[36]

The handle is composed of the *spindle*, in ash or beech, wrapped in cord. The guard, of iron, presents two *lens*.[37]

The Manner of Holding the Sword

The handle of the sword in the right hand, the thumb extended above and almost touching the lens, the four other fingers together underneath.

Hold the sword with the thumb and index finger only, the other fingers are pushed constantly on the handle and only squeeze it in order to support the parry in the moment when one executes it. Squeezing the sword constantly forces [one] to execute with the wrist and not with the fingers.

Fig. 1

The Guard

The guard is the preliminary position indicated by experience as the best to take by the fencer to be ready for both attack and defense.

The guard is taken in seven steps,[38] after the execution of certain preparatory movements. At the start of instruction, one has always to follow the *mise en guard*:[39] the gathering, the step, the *appels*, the deployment of the arms and the lunge, re-taking the guard and the salute of arms.

Preparatory Movements

Being in the stance of a soldier without weapons, make a half-left [turn], maintain a straight head, the feet placed squarely without separating[40] the heels, the right arm extended in front and separated from the body, the point of the sword around 10 centimetres from the ground, the left arm falling naturally, the hand open.

Taking Guard in Seven Steps

1. Lift the sword, the hand to the height of the eyes, the thumb above, the arm extended, the sword in the extension of the arm (fig. 2).

2. Lower the sword, the arm straight, the point around 10 centimetres from the ground (fig. 1).

3. Bring the sword again horizontally against the body turning the fingers of the right hand below, bringing at the same time the left hand bent against the guard, the fingers extended and touching the guard, the top of the fingers on the blade (fig. 3 and 4.).

4. Lift the sword bending the arms and passing them close to the body, placing it horizontally above the head, the arms extended.

---

[32] *monture*
[33] *mouche*
[34] *soie*
[35] *poignet*
[36] *garde*
[37] *lunettes*
[38] *en sept temps*
[39] lit: the putting into guard
[40] *désunir*

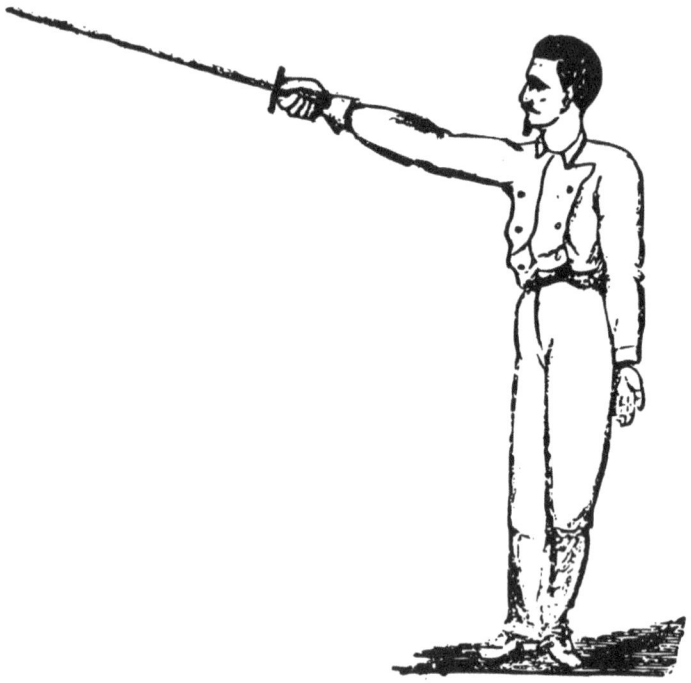

Fig. 2

Fig. 3 and Fig. 4

Fig. 5

Fig. 6

5. Flex the arms towards the head, carrying the left hand backwards at the height of the head, the hand curved, the thumb lightly separated from the other fingers, the arm rounded, lower the right hand bending the elbow[41] and placing it, the thumb above and at the height of the right breast, the elbow on the inside, around 15 centimentres before the body, the point of the sword at the height of and between the two eyes (fig. 5).

6. Flex the legs spreading the two knees, the body plumb on the hips (fig. 6).

7. Maintain the body's weight on the left leg, extend the right leg to its full length, placing the foot flat on the ground in front and before the left heel. Flex lightly on the left leg in order to advance the right knee so that it becomes perpendicular to the middle of the foot (fig. 7).

---

[41] *en ployant la saignée du bras*, lit: bending the bleeding of the arm?

Fig. 7

Gathering

The gathering[42] is the return to the standing position, being that of the guard.

To execute it forward

Lift the right arm in front, the hand to the height of the eyes, and let the left hand fall in line. Then return the rear heel against the front heel, straightening up.

To execute it backwards

Lift the arm, the hand to the height of the eyes, and let the left hand fall in line. Then bring back the front heel against the rear heel straightening up.

Gathering is employed after each *reprise*, whether for granting a momentary rest and to allow continuing the lesson, or for ceasing all work. In this latter case, after having gathered, execute the salutes and order "rest".

Stepping

Forward

Carry the right foot forward without disarranging either the position of the body or that of the sword, following immediately [with] the left foot to its distance.[43]

Backward

Carry the left foot backwards without disarranging either the position of the body or that of the sword bringing the right foot immediately backwards to its distance.

Appels

Lightly hit the ground two times in a row with the right foot, the body remaining immobile.

Be able to execute them at whatever moment without being obliged to move the body.

---

[42]*rassemblement*
[43]position?

Deployment of the Arm

Extend the right arm without jerking and without acting from the shoulder, the body remaining immobile, turning at the same time the right hand, the fingers lightly upwards, and placing it, as well as the point of the sword, at the height of the chin (fig. 8).

Bend[44] the right forearm, equally without jerking, in order to return[45] to first position.

Maintain the mounting and do not let it go in order to place more easily the sword horizontally, but make an act of wrist articulation turning the hand slightly, fingers upwards.

Look always over the wrist.

Lunge

The lunge is the development by which the fencer impresses the sword into the most vigorous and extended action.

It is executed in the following fashion: extend the right arm as prescribed above, straighten briskly the left knee, carry at the same time the right foot forward and in front of the left heel, the foot passing close to the ground. Place the foot flat in a manner that the knee is outside and perpendicular to the middle of the foot, the body plumb and following the impulse of the same leg, letting the left arm fall behind at the same time along and 10 centimentre from the thigh, the fingers of the hand extended and joined, the thumb separate, the head straight, the eyes fixed on the point of the sword (fig. 9).

In order to retake the first position: Nimbly bend the left knee, lifting the left arm, the weight of the body carried on the left foot, shorten the right arm and carry at the same time the right foot to its position, the foot passing close to the ground, placing it flat without striking.

Ensure the body is plumb by raising it, by the *appels*.

Salute in Arms

After having executed the Gather:

Salute Before You

1. Shorten the arm, the elbow joined to the body, the hand at the height of the chin, the fingers turned towards the body.

---

[44] *ployer*
[45] *reprendre*

Fig. 8

Fig. 9

2. Lower the blade extending the arm, the right hand the fingers above and to the side of the right thigh.

The Lines (fig. 10)

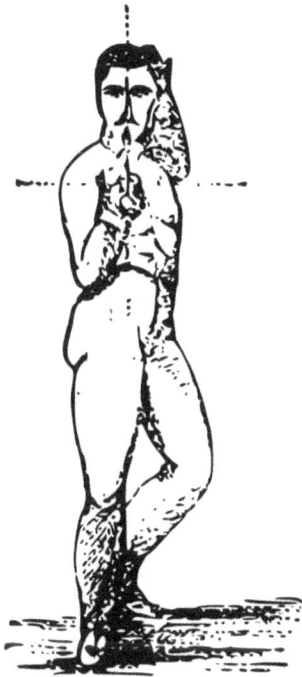

The lines are the zones or parts of space into which the sword moves leaving the guard position.

They are four in number:

The right line or *tierce* or outside, which is the zone of part of space to the right of the sword.

The left line or *quarte* or inside, which is the zone or part of space to the left of the sword.

The high line or *prime*, which is the zone or part of space above the wrist.

The low line or *second*, which is the zone or part of space below the wrist.

The right line and the left line are the only lines of engagement.

The high line and the low line are the the lines into which one strikes.[46]

The high line beats[47] the low line.

Fig. 10

Only strike in the low line (*dérobement*) after having closed the high line by a feint, a beat or an opposition,[48] in a manner not to be stopped by a timed strike.

Engagement

Engagement is the joining of the opponent's steel from the opposite side where one had first made contact in order to cover oneself.

[46] *dans lesquelles on tire*
[47] *primé*
[48] *pression*

Fig. 11 and Fig. 12

In order to execute it:

Join the steel, carrying the wrist, the thumb above, to the left (fig. 11) (or to the right) (fig. 12) in order to cover oneself.

Having thus made contact, in order to make an engagement: lower the point of the sword, passing it by the shortest line below the opposed sword. Join the steel carrying the wrist, the thumb above, to the right (or to the left) in order to cover oneself, and place the point of the sword at the height and opposite the eye on the side of the engagement.

Change of Engagement

The change of engagement is a new engagement made on the opposite side of the previous.

In order to execute it:

On the end of the first engagement, lower the point of the sword, passing it by the shortest line below the opposed sword. Join the steel carrying the wrist, the thumb above, to the right (or to the left) in order to cover oneself and place the point of the sword at the height of and opposite the eye on the side of the new engagement.

The engagement and change of engagement take place on the firm foot and stepping.

Joining the steel alone must always been done on the firm foot and precede the execution, whether from an engagement or from a blow struck.[49]

Double Engagement

The double enegagement is the immediate succession of two engagements.

In order to execute it:

Make immediately and without disarrnging the wrist two engagements, the first beginning on the line opposed to that in which one finds oneself.

The double engagement takes place on the firm foot or stepping. In this last case, the step must be finished at the moment of the execution of the second engagement.

Execute with the fingers,[50] without disarranging the wrist, the second engagement serving as a parry to the first.

---

[49] *d'un coup porté*
[50] *S'attacher à executer les doigts*

Fingering

Fingering[51] is the momentary displacement of the sword's point by the action of the fingers, particularly with the thumb and the first two fingers, which determines a feints, directing the point in the attack and helping to contain the opposing sword in the parry.

Fingering gives the fencer the quality of delicacy and finesse.[52]

Attack

The attack is the action of the fencer which seeks to hit his adversary, carrying to him a simple or compound strike.

The strike is simple when it results from only a single movement. It is compound if it follows[53] from several (2, 3, or 4).

The simple strike consists of three varieties: the straight strike,[54] the disengagement and the cutover.[55]

Straight Strike

The straight strike hits the adversary directly. It results from the deployment of the arm and from the lunge.

In order to execute it:

The line of engagement not being firm, deploy the arm, lunge as has been described, and touch without releasing the handle, the hand elevated and maintaining the line in order to cover oneself.

Disengagement

The disengagement is a change of lateral line, following the straight strike.

In order to execute it:

Lower the point of the sword, making only the fingers act, passing it by the shortest line below the opposing sword, deploying the arm, the wrist before one, lunge and touch. The wrist is carried to the side of the sword in order to cover oneself.

---

[51] *doigté*
[52] *ténuité*
[53] *découler*
[54] *le coup droit*
[55] *coupé*

Feint

The feint is the pretense of a strike. It must be executed in such a manner that, taken for the strike itself, it necessitates a parry. One must, in consequence, execute it holding oneself ready to deceive the parry [the opponent] makes, either by another feint or by a real strike.

The feint is an ordinary strike executed like the strike itself but without lunging.

Cut-over

The cutover is just a disengagement over the point.

It is executed in the same manner but raising the point of the sword, the fingers alone acting, in order to pass it the closest possible above the point of the opposing sword, deploying the arm.

Glide

The glide is the feint of the straight strike, gliding softly along the sword.

Beat

The beat is a more or less light shock from the point to the point in order to unsettle the opposing sword and carry a strike more easily.

Pressing

Pressing is pushing more or less lightly from point to point, in order to unsettle the opposing sword and carry a strike more easily.

Expulsion

The expulsion[56] is but a sudden pressing, prolonged and sliding.

Disengagement

The disengagement is the attack carried out[57] favouring the passage of the sword from the high line to the low line.

In order to execute it:

---

[56] *froissement*
[57] *opérée*

After a feint, beat or pressing, lower the point of the sword in order free [it], lunging and touching in the low line on the same side, raising the wrist as high as possible.

Bind

The bind[58] is the action by which one captures the opponent's sword, pushing on the weak of that sword with the strong of one's own, in order to take it from the high line into a low line, and vice-versa.

In order to execute it:

Make a parry of eight (or semicircle, following the engagement),[59] pushing with the strong on the weak of the opposing sword and deploying the arm. Lunge and touch in the low line without ceasing to hold the [opponent's?] sword.

The feint of the bind is executed like the bind but without the lunge.

The bind can only be executed on the extended arm, above all when in this position one meets strong resistence.

Remise

The *remise* is an attack executed on the absense or abandonment of the sword after a parry. It is produced after a parried attack and the abandonment of the sword, made in order to riposte, putting the point again on line without being lifted, and hitting with a straight strike. The intelligent *remise* is to the riposte that which tempo is to the attack.

Reprise

The *reprise* is the renewing of the attack after having retaken the sword without being raised, on a parry without riposte. It is produced after a parried attack and not following the riposte, retaking the attack, in raising it and trying to strike.

Redoubling

Redoubling is the immediate succession of two attacks without being raised, on a parry without riposte. It is produced after a parried attack

---

[58] *liement*

[59] *la parade d'octave*: blade down and to the outside, wrist supinated, the point lower than the hilt

and not following a riposte, retaking the attack immediately, without being raised, and trying to strike.

Timed Strike

The timed strike[60] is an attack surprising the opponent in the preparation of his own. It is therefore executed in the absence of the sword, a too large feint or an attack directed against the low line, or an attack compromised[61] by the movement of the foot before the deployment of the arm.

It is produced on a feint outside the body or a direct attack in the low line, deploying the arm in the high line without seeking the sword and touching with straight strike.

The timed strike is thus a single movement which forms at the same time a parry and riposte. It is an indivisible parry and riposte. It consists, in summary, of preventing the adversary in the final execution of his compound attack, closing the line where he will hit. It is preferable to parry in the high line than in the inside line, because thus it is less exposed to the double hit. After having familiarised the student with the engagements, attacks, parries and ripostes, it is good to introduce him to different timed strikes.

Stop Strike

The stop strike is a lively attack executed on another passing attack, preceded by many feints. It is a timed strike taken on the adversary's step.

Parry

The parry is the action of turning away from the body a blow struck. It is executed always with the strong on the weak and chases the sword, casting it aside in the same line in which it is presented.

Counter

The counter is an inverse parry which seeks the sword in the line in which it is presented in order to bring it and chase it in the opposing line. It is the circular parry.

---

[60] *coup de temps*
[61] *décomposée*

The Actual Parry

The parry properly named[62] is that which chases the sword without accompanying [it]; that is to say, suddenly turning it away from the body (in order to facilitate a riposte) by a sharp beat[63] made with the aid of the fingers.

Opposition

The opposition is a special parry which chases the sword accompanying it; that is to say, turning it away from the body without jerking [it] and by a single action of the wrist.

Varieties of Parry

There are eight parries, each having its counter. These are: *prime, seconde, tierce, quarte, quinte, sixte*, the half-circle or *septime*, and *octave*.

The parries are executed in the following manner:

The parries of *seconde* and *octave* take place to the right in the low line.

The parries of *tierce* and *sixte* take place to the right in the high line.

The parries of *quarte* and *prime* take place to the left in the high line.

The parries of *quinte* and of the half-circle or *septime* take place to the left in the low line.

The Execution of *Prime*

On a strike thrown in the left line, turn the hand the finger before,[64] lift the elbow bending it[65] and place the forearm horizontally, the wrist above the left eye, the point of the sword threatening the low line (fig. 13).

Being in the above position, in order to make the counter of *prime*, lift the point of the sword and bring it into the above position passing above the opposing sword.

Fig. 13

Fig. 14

The Execution of *Seconde*

On a strike thrown in the low line, lower the wrist and impress on it a movement from left to right, turning the hand fingers downwards, the forearm, the wrist and the sword in a horizontal position.

Being in the above position, in order to make the counter of *seconde* on a strike thrown in the low line, lift the point of the sword and bring it to the first position passing above the opposing sword (fig. 14).

The Execution of *Tierce*

On a strike thrown in the right line, carry the hand to the right the fingers below, making an articulation of the wrist and placing the point of the sword at the height and facing the left eye (fig. 15).

Being in the above position, in order to make the counter of *tierce*, on a strike thrown in the left line, lower the point of the sword and bring it to the first position passing below the opposing sword.

Execution of *Quarte*

On a strike thrown in the left line, carry the hand to the left, the thumb lightly to the right, having the articularion of the wrist act lightly and place the point of the sword at the height of and facing the left eye (fig. 16).

Being in the above position, in order to make the counter of *quarte*, on a strike thrown in the right line, lower the point of the sword and bring it to the first position passing below the opposing sword.

Execution of *Quinte*

On a strike thrown in the left line, turn the hand slightly, fingers upwards, lower the wrist crossing the opposing sword in order to beat it down into the low line, the forearm, the wrist and the sword in a horizontal position and perpendicular in respect to the body (fig. 17).

Being in the above position, in order to make the counter of *quinte*, on a strike thrown in the right line, lift the point of the sword and bring it the first position passing above the opposing sword.

---

[62] *la parade proprement dite*
[63] *un battement sec*
[64] *en avant*
[65] *la saignée - ?*

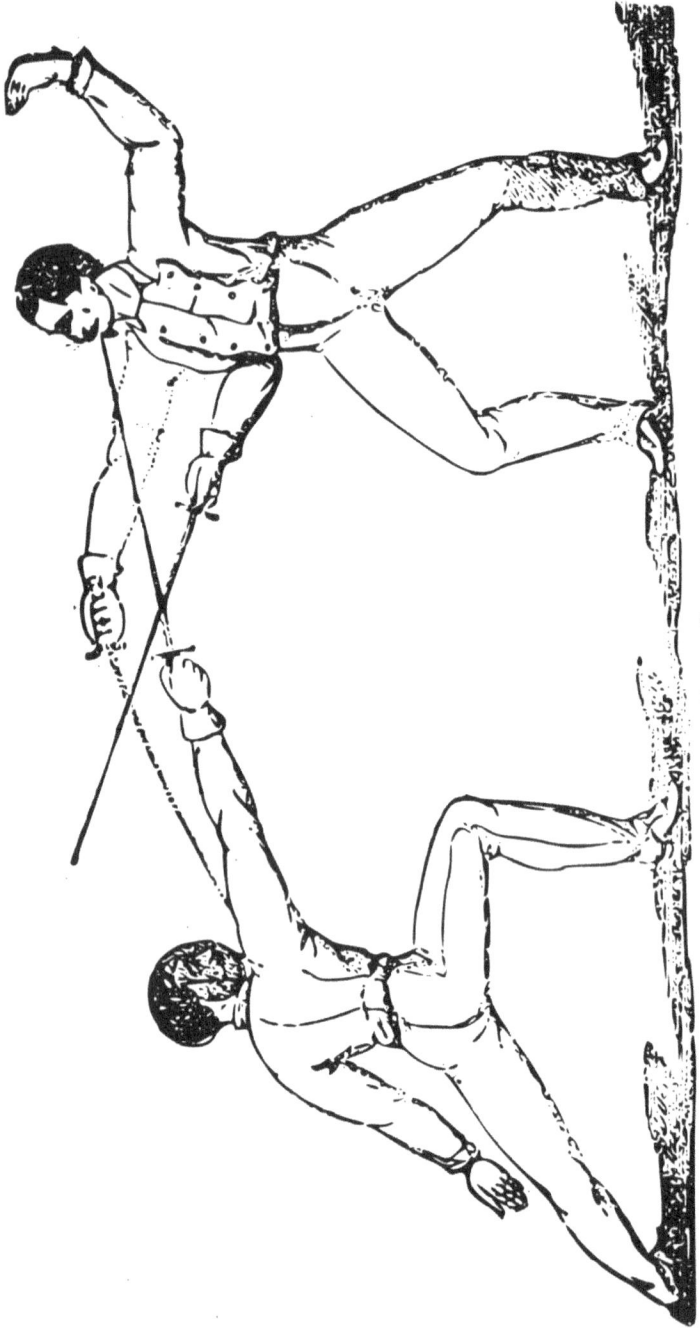

Fig. 15

Fig. 16

Fig. 17

Execution of *Sixte*

On a strike thrown in the right line, carry the hand to the right, the thumb to the side, the fingers slightly upwards, making an articulation of the wrist and place the point of the sword at the height of and facing the right eye (fig. 18).

Being in the above position, in order to make the counter of *sixte* on a strike thrown in the left line, lower the point of the sword and bring it to the first position passing below the opposing sword.

Execution of the Half-Circle

On a strike thrown in the low line, lower the point of the sword and place it below the opposing wrist, turning the hand the fingers upwards, the wrist executing a movement from right to left, remaining always at the same height (fig. 19).

Being in the above position, in order to make the counter of the half-circle on a strike thrown in the low line, lift the point of the sword and bring it to the first position passing above the opposing sword.

Execution of *Octave*

On a strike thrown in the low line, lower the point of the sword and place it below the opposing wrist, turning the fingers slightly upwards, the wrist remaining at the same height (fig. 20).

Being in the above position, in order to make the counter of the *octave* on a strike thrown in the low line, lift the point of the sword and bring it to the first position passing above the opposing sword.

Riposte

The riposte is an attack which follows the parry, either immediately or after a moment determined by the movements of the opponent. The riposte which immediately follows the parry is called the "tick-tack" riposte,[66] that is to say, a riposte in which the "tick" (the action of touching the sword) is followed immediately by the "tack" on the chest.

Counter-riposte

The counter-riposte is an attack which follows immediately the parry of a riposte.

---

[66] *la riposte du tac au tac*

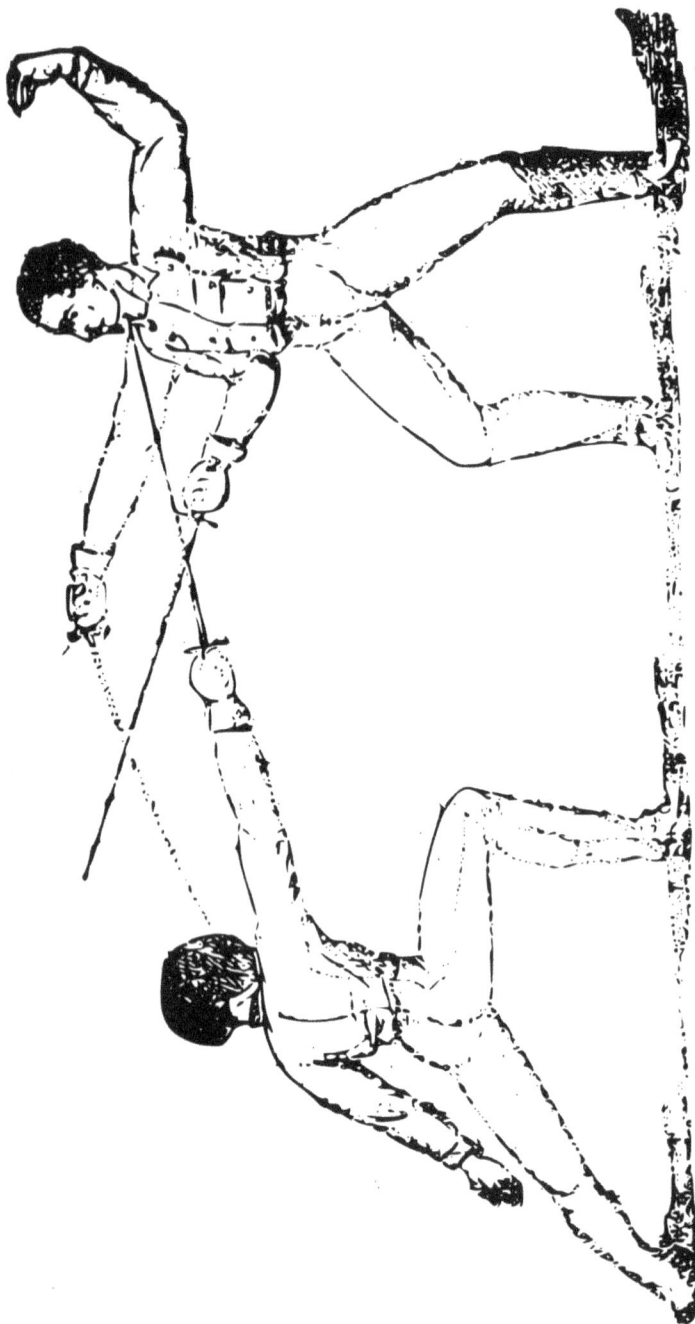

Fig. 18

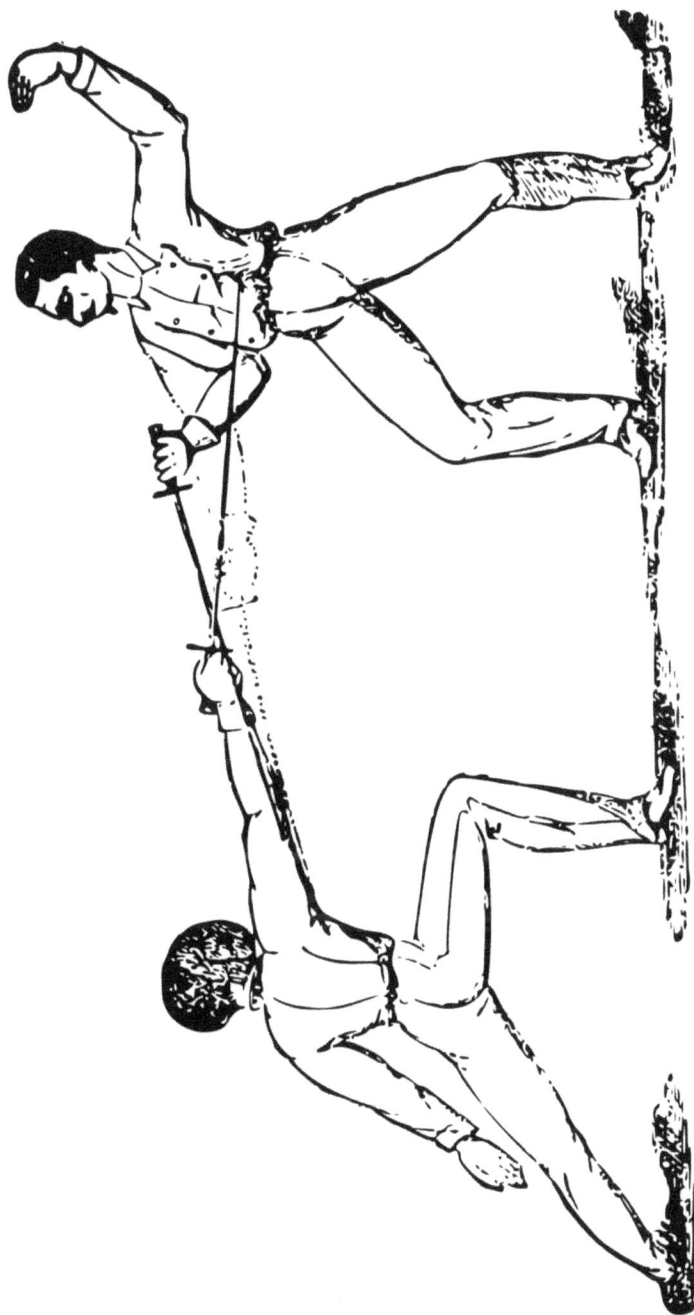

Fig. 19

Fig. 20

The Fencing Phrase

The fencing phrase[67] is the sequence of several strikes made and received[68] without interruption.

Article III — Learning Progression

Instruction is done at the *plastron* in seven lessons and three iterations of each.

First Lesson

1. Nomenclature of the sword. The manner of holding the sword. Preparatory movements. Taking guard. The salute at arms.

2. Exercises on the direct attacks. Parries on disengagement and ripostes with the straight strike.

3. Exercises on the direct attacks. Parries with two disengagements and ripostes with the straight strike.

Second lesson

1. Exercises on various attacks (beats, pressings, explusions). Parries with three disengagement and riposte with the straight strike.

2. Exercises on various attacks (the cutover, the glide). Parries on four disengagements and ripostes with the straight strike.

3. Exercises on various atatcks in the low line (the glide, beat and pressing). Parries on one, two, three disengagements and ripostes with the staright strike.

Third Lesson

1. Exercises on changes of engagement and on compound ripostes.

2. Exercises on the double engagement and on compound ripostes.

3. Exercises on stepping (attacks, parries and ripostes with the straight strike).

---

[67] *la phrase d'armes*
[68] *portés et rendus*

Fourth Lesson

1. Exercises on stepping and the change of engagement and various simple ripostes.

2. Exercises on stepping and the double engagement and various simple ripostes.

3. Exercises on stepping, double engagement and the change of engagement and compound ripostes.

Fifth Lesson

1. Exercises on the beat in the line opposed to the engagement and on various parries.

2. Exercises in the low line (stepping and double engagement) and ripostes with the straight strike.

3. Exercises to deceive the double engagement and on the absence of the sword in attack and in riposte.

Sixth Lesson

1. Exercises on the simple counter-riposte after various parries.

2. Exercises on the simple counter-riposte and the compound counter-riposte.

3. Exercises on parrying the counter-riposte.

Seventh Lesson

1. Exercises on direct and compound attacks of one or two disengagements. Parries: first by oppositions and ripostes with the straight strike; second by counters and ripostes with the straight strike.

2. Exercises on direct and compound attacks of two or three disengagements. Parries: first by oppositions and counters and ripostes with the straight strike; second by counters and oppositions and ripostes with the straight strike; third by counters and oppositions and counters and ripostes with the straight strike.

3. Exercises on direct and compound attacks of one or two disengagements preceded by a double disengagement. Parries: first by oppositions and ripostes with the straight strike; second by counters and ripostes with the straight strike.

Article IV — General Rules to Observe

1. Execute slowly at first in order to give the time to understand the blow struck. Endeavour to acquire speed later - but with the sword, not with the voice.[69]

2. Follow each iteration in the left line first and then in the right line.

3. Precede each exercise with the command, "Ready swords."[70]

4. Complete each exercise with a lunge executed at the command, "Lunge"[71] (or without command as soon as the enunciation of the final syllable of the strike to make,[72]) and by coming again on guard at the command, "*En garde*" (or without command as the strike has succeeded or has not touched).

5. End each exercise with the straight strike, lunging after the parry, always made in anticipation of the exercise which follows.

6. Rest during the reprise, gather [and] immediately retake guard as long as the lesson has not finished.

7. End the lesson with the command: first "Gather forward", second "Salute in front".

Article V — Details of the Lessons

First Lesson

First Iteration

Sword nomenclature.
The manner of holding the sword.
Preparatory movements.

---

[69] *et non avec la voix*
[70] *Engagez l'épée*
[71] *Fendez-vous!*
[72] *du coup à porter*

Going on guard

1. On guard in seven steps
2. *En garde*

Assemble

1. Gather forward (or backwards)

Lunge

1. Deploy the arm ⎫
2. Lunge ⎬ Do not touch the *plastron*
3. *En garde* ⎭

Fingering exercises

1. Fingering exercise
2. On the parry of *quarte* (or *tierce*, half-circle, *seconde*, *sixte*, *octave*).

Note: Sometimes avoid the parry in order to assure that the hand remains in line. End with a lunge.

Engagements and Definitions

Engagement

Engage the sword. Touch the plastron with a lunge.

Note: Repeat following the engagement in two lines several times. Following, add stepping, *appels* and the lunge in order to provide balance.[73]

Definition of the Lines

Four lines: left, right, high, low.

Definitions of the Parries

Eight Parries
To the left:
High: *quarte, prime*
Low: half-circle, *quinte*

To the right:

---

[73] *afin de donner l'assiette*

High: *sixte, tierce*
Low: *octave, seconde*

Note: demonstrate by executing them.

Divisions of the Blade

The point: the weak part with which one strikes
The heel: the strong part with which one parries
The middle: the intermediate part
Engagements must be taken in the interval between this part and the fly or button.

Salute At Arms

Salute in front.

Second Iteration

Direct attacks

The Straight Strike

1. Engage the sword
2. Deploy the arm ⎫ Or, without   ⎫ Have it executed
3. Lunge   ⎬ deconstructing,   ⎬ with and without
4. *En garde*   ⎭ strike straight   ⎭ deconstructing

   Note: Repeat this exercise several times, allowing it to touch each time. To end, parry with respect to the following exercise and have it touch with a straight strike without engaging again. This direction is common to all the exercises.

Disengaging

2. Disengage, remain (or retire)

Doubling

2. Feint a disengagement
3. I make the counter: deceive[74] (or double)
4. Lunge

---

[74] *trompez*

Doubling and Disengaging

2. Feint a disengagement
3. I make the counter and two oppositions : deceive, disengage (or double, disengage)

Doubling and One, Two

2. Feint a disengagement
3. I make the counter and two oppositions: deceive, one, two (or double, redouble)

Execute one of the two strikes according to the third parry.

Doubling in the Two Lines

3. I make the counter, an opposition and a counter: deceive, double (or double, redouble)

Change the line

Parries and Ripostes of the Straight Strike

On a disengagement:

1. Engage the sword
2. *Sixte*, oppose (or parry) and riposte straight — After an opposition, parry with an opposition
3. *En garde*

2. Counter of *quarte*, oppose (or parry) and riposte straight
2. *Quarte*, oppose (or parry) and riposte straight — After a counter, parry with a counter
2. Counter of *sixte*, oppose (or parry) and riposte straight

Third Iteration

Direct attacks

1. Engage the sword
2. Feint a disengagement
3. I make the counter: disengage — or one, two
4. Lunge
5. *En garde*

One, Two, Deceive the Counter

2. Feint a disengagement
3. I make an opposition, a counter, double (or one, two, deceive)

One, Two, Deceive the Counter and Disengage

2. Feint a disengagement
3. I make an opposition, a counter, an opposition: double, disengage (or one, two, deceive, disengage)

One, Two, Three

2. Feint a disengagement
3. I make two oppositions: one, two. (or one, two, three)

One, Two, Three, Deceive the Counter

2. Feint a disengagement
3. I make two oppositions and a counter: one, two, deceive (or one, two, three, deceive)

Execute one of the two strikes according to the third parry.

One, Two, Three, Four

3. I make three oppositions: one, two, three (or one, two, deceive)

Change the line

Parries and Ripostes of the Straight Strike

On Two Disengagements: First: One, Two

2. *Sixte* and *quarte*: oppose (or parry) and riposte

3. *Sixte* and counter: oppose (or parry) and riposte

Change the line

Second: Doubling

2. Counter of *quarte* and *sixte*: oppose (or parry) and riposte

3. Counter of *quarte* twice: oppose (or parry) and riposte

Change the line

Second Lesson

First Iteration

Diverse attacks

Beat, Straight Strike

2. Beat, strike straight

Beat, Disengagement

2. Beat, disengage

Beat, Doubling

2. Beat, double

Beat, Doubling, Disengagement

2. Beat, double, disengage

Beat, One, Two

2. Beat, one, two

Beat, One, Two, Deceive

2. Beat, one, two, deceive

Beat, One, Two, Three

2. Beat, one, two, three

Change the line

Pressing, Straight Strike

2. Press, strike straight

Pressing, Disengagement

2. Press, disengage

Pressing, Doubling

    2. Press, double

Pressing, Doubling, Disengagement

    2. Press, double, disengage

Pressing, One, Two

    2. Press, one, two

Pressing, One, Two, Deceive

    2. Press, one, two, deceive

Pressing, One, Two, Three

    2. Press, one, two, three

   Change the line

Gliding, Strike Straight

    2. Glide, strike straight

Gliding, Disengagement

    2. Glide, disengage

Gliding, Doubling

    2. Glide, double

Gliding, One, Two

    2. Glide, one, two

Change the line

Parries and Ripostes of the Straight Strike
   On three disengagements:

First: One, Two, Three

2. *Sixte*, *quarte*, *sixte*: oppose (or parry) and riposte

3. *Sixte*, *quarte*, counter: oppose (or parry) and riposte

Second: Doubling, Disengagement

2. Counter of *quarte*, *sixte*, *quarte*: oppose (or parry) and riposte

3. Counter of *quarte*, *sixte*, counter: oppose (or parry) and riposte

Third: One, Two, Deceive the Counter

2. *Sixte*, counter, *quarte*: oppose (or parry) and riposte

3. *Sixte*, counter twice: oppose (or parry) and riposte

Change the line

Second Iteration

Diverse attacks

Cutover[75]

2. Cutover: remain (or: cutover)

Circling the Sword

2. Feint a cutover: disengage

Circling the Sword, Disengagement

2. Feint a cutover, one, two

Change the line

---

[75] *Coupé*

Thrust with Opposition[76]

2. Feint a straight strike, strike straight. Parry in Opposition

Thrust with opposition, disengagement. Parry with a Counter.
Thrust with opposition, doubling.
Thrust with opposition, one, two
Thrust with opposition, deceive the counter
Thrust with opposition, deceive, disengage
Thrust with opposition, deceive twice
Thrust with opposition, cutover
Thrust with opposition, circle the sword
Change the line

Parries and Ripostes of the Staright Strike

On four disengagements:

First: One, Two, Deceive, Disengage

2. *Sixte*, counter, *quarte, sixte*: oppose (or parry) and riposte

3. *Sixte*, counter *quarte*, counter: oppose (or parry) and riposte

Change the line

Second: Doubling, One, Two

2. Counter of *quarte, sixte, quarte, sixte*: oppose (or parry) and riposte

3. Counter of *quarte, sixte, quarte*, counter: oppose (or parry) and riposte

Change the line

Third: One, Two, Three, Deceive

2. *Sixte, quarte*, counter, *sixte*: oppose (or parry) and riposte

3. *Sixte, quarte*, counter twice: oppose (or parry) and riposte

Change the line

---

[76] *coulé*

Fourth: Doubling in Two Lines

2. Counter of *quarte*, *sixte*, counter, *quarte*: oppose (or parry) and riposte

3. Counter of *quarte*, *sixte*, counter twice: oppose (or parry) and riposte

Change the line

Third Iteration

Attacks in the low line

Disengagement in Time[77]

2. Feint a straight strike
   beat (or parry)
3. Strike below
4. *En garde* } Oppose [with] half-circle or *quinte*

Disengagement in Time Deceiving the Half-Circle

1. Feint a straight strike, disengage in time
2. Feint a straight strike, feint below, deceive the half-circle
3. Feint a straight strike, feint below, strike above
4. Feint a straight strike, feint below, one, two above

Change the line

Binding[78]

2. Bind, remain (or bind), parry in *octave* or the half-circle

Binding, Strike Above

2. Feint a bind
3. Strike above } Or bind, strike
above } Parry with
opposition in *sixte*

---

[77] *dérobement*
[78] *liement*

Bind, One, Two Above

2. Feint a bind   } Or bind, one,   } Change
3. One, two above } two above   } the line

Parries in the Low Line and Ripostes from the Straight Strike

On one disengagement:

    2. Half-circle (or *octave*): oppose (or parry) and riposte

On two disengagements:

First, in the low line

    2. Half-circle, *octave*: oppose (or parry) and riposte

    3. Half-circle, counter: oppose (or parry) and riposte

    Change the line

Second, in the low line and the high line

    2. Half-circle, *quarte*: oppose (or parry) and riposte

    Change the line

Third, in the high line and low line

    2. *Sixte, octave*: oppose (or parry) and riposte.

    Change the line.

    2. Counter of *quarte*, half-circle: oppose (or parry) and riposte.

    Change the line.

On three disengagements:

First in the high line and twice in the low line

2. *Sixte*, *octave*, half-circle: oppose (or parry) and riposte

3. *Sixte*, counter, half-circle: oppose (or parry) and riposte

4. Counter of *quarte*, half-circle, *octave*: oppose (or parry) and riposte

5. Counter of *quarte*, half-circle, counter: oppose (or parry) and riposte

Change the line

Second in the low line and twice in the high line

2. Half-circle, *quarte*, *sixte*: oppose (or parry) and riposte

3. Half-circle, *quarte*, counter: oppose (or parry) and riposte

Change the line

Third Lesson

First Iteration

Attacks deceiving the change of engagement

Straight Strike

1. Engage the sword
2. At the change of engagement, deploy the arm or without deconstructing, strike straight
3. Lunge
4. *En garde*

Disengagement
Doubling
Doubling, disengagement
One, two
One, two, deceive the counter
One, two, three
Beat, straight strike
Beat, disengagement
Beat, one, two

Change the line

Parries after changing and compound ripostes
On one or two disengagements:

Straight Strike

1. Engage the sword, deceive sometime the engagement
2. *Sixte*: oppose (or parry) and riposte straight
3. *En garde*

Riposte withdrawing[79]
Riposte doubling
Riposte with a one, two
Riposte with a cutover
Riposte circling the sword
Change the line
One opposition
A counter
Two oppositions
One opposition and a counter
A counter and an opposition
Two counters

Second Iteration

Attacks preceded by a Double Engagement

Straight Strike

1. Double the engagement

2. Deploy the arm ⎫ Or without
3. Lunge   ⎬ deconstructing
4. *En garde* ⎭ strike straight

Disengagement
Doubling
Doubling, disengagement
One, two
One, two, deceive the counter
One, two, three

---

[79] *ripostez en dégageant*

Cutover
Circling the sword
Change the line

Parries after a double engagement and compound ripostes

On one or two disengagements:

Straight Strike

1. Double the engagement (deceive sometimes)
2. *Sixte*: oppose (or parry) and riposte straight
3. *En garde*

Riposte withdrawing
Riposte doubling
Riposte with a one, two
Riposte with a cutover
Riposte circling the sword
Change the line

Note: Vary the parries in the order determined by the first itera-tion above.

Third Iteration

Stepping attacks (or at the step)

Straight Strike

1. Engage the sword, advance (or withdraw)
2. Deploy the arm (or at
   the step, deploy the arm) ⎫
3. Lunge ⎬ or strike straight
   ⎭
4. *En garde*

Disengagement
Doubling
Doubling, disengagement
One, two
One, two, deceive the counter
One, two, three
Cutover
Circling the sword

Change the line

In the compound strikes, execute the first feint immediately on the step.

Parries after the step (or an attack advancing), without (or with) double engagement, and riposte with the straight strike.

On one disengagement:

1. Engage the sword (or double the engagement)
2. *Sixte*: oppose (or parry) and riposte straight
3. *En garde*

Counter of *quarte*: oppose (or parry) and riposte straight
Quarte: oppose (or parry) and riposte straight
Counter of *sixte*: oppose (or parry) and riposte straight

Fourth Lesson

The fourth and fifth lessons must be considered as sparring lessons.[80]

First Iteration

Stepping attacks (or on the step) deceiving the change of engagement

Straight Strike

1. Engage the sword (advancing or withdrawing)
2. At the change of engagement (or on the step and at the change of engagement), deploy the arm or without deconstructing, strike straight
3. Lunge
4. *En garde*

Disengagement
Doubling
One, two
Beat, straight strike
Beat, disengagement
Beat, one, two
Change the line

Parries after the step (or after a stepping attack) and various simple ripostes

---

[80] *leçons d'assaut*

On one or two disengagements:

1.
  1. Engage the sword, advance (or withdraw).
     Deceive the engagement sometimes
  2. *Sixte*: oppose (or parry) and riposte above and below
  3. *En garde*

2.   2. Riposte with the point by a cutover above and below
3.   2. Riposte with the point by a circling of the sword
4.   Cutover with the point, parry in *quarte* and riposte

Change the line
One opposition
A counter
Two oppositions
An opposition and a counter
Two counters

Note: After the *sixte* and the counter, make a riposte above (or below) hand inverted in *prime*. Make varied parries in the order above.

Second Iteration

Stepping attacks (or on the step) preceded by a double engagement.

Straight strike

1. Double the engagement: advance (or withdraw)
2. Deploy the arm (or on
   the step, deploy the arm)   Or without decomposing
3. Lunge   strike straight
4. *En garde*

Disengagement
Doubling
One, two
Cutover
Circling the sword
Change the line

Parries after the step (or a stepping attack) preceded by a double engagement and simple and varied ripostes on one or two engagements.

1. Double the engagement, advance (or withdraw). Deceive the double engagement sometimes.

2. *Sixte*: oppose (or parry) and riposte above or below
3. *En garde*

Riposte with the point by a cutover above or below
Riposte with the point by a circling of the sword.
Cutover with the point, parry in *quarte* and riposte.
Change the line.
Note: Conform to that which is prescribed in the first iteration's note above.

Third Iteration

Stepping attacks preceded by a double engagement and deceiving the change of engagement

Straight Strike

1. Double the engagement: advance (or withdraw)
2. At the change of engagement (or on the step and change of engagement) deploy the arm) } Or without deconstructing strike straight
3. Lunge
4. *En garde*

Disengagement
Doubling
One, two
Beat, straight strike
Beat, disengagement
Beat, one, two
Change the line

Parries after the step (or a stepping attack) preceded by a dopuble disengagement and compiound ripostes.

On one or two disengagements:

Straight strike

1. Double the engagement: advance (or withdraw)
2. *Sixte*: oppose (or parry) and riposte straight

Riposte withdrawing
Riposte doubling
Riposte with a one, two
Riposte with a cutover
Riposte circling the sword
Change the line
Note: Vary the parries in the order determined in the first iteration above.

Fifth Lesson

First Iteration

Attack by a beat in the line opposed to the engagement.

Beat, Straight Strike

1. Engage the sword
2. In changing the engagement: beat
3. Deploy the arm ⎱ Or without deconstructing:
4. Lunge   ⎰ strike straight
5. *En garde*

Beat, disengage
Beat, double
Beat, one, two
Change the line
Various parries and simple ripostes

On the disengement:

1. Engage the sword (left line)
2. Parry in *tierce* and (hand in *quarte* or *prime*)
3. En garde

Oppose in *sixte* and riposte below with a *croisé*.[81]
Take a half-circle underneath the sword.
Parry with the counter of *tierce* and riposte as above. Engage in *tierce*.
Oppose in *seconde* and riposte.
Parry with *seconde* and riposte (above the hand in *quarte*).

---

[81] The term is the same in English: an action which carried the opponent's blade from high to low on the same side.

Parry in *quinte* and riposte

Oppose *quarte* and riposte (below) with a *croisé*. Take an *octave* underneath the sword.

Parry with *prime* and riposte below.

Parry with *prime* and riposte with a cutover.

Parry with *prime*, cutover, parry in *quarte* or *quinte* and riposte.

Second Iteration

Attacks in the low line stepping without (or with) double engagement

Glide and Disengage in Time[82]

1. Engage the sword (or double engagement)

2. Feint a straight strike ⎫ Or without deconstructing:
3. Strike below   ⎭ cutover, disengage in time

4. *En garde*

Beat, disengage

Beat, double

Beat, one, two

Change the line

Various parries and simple ripostes

On the disengement:

1. Engage the sword (left line)
2. Parry in *tierce* and (hand in *quarte* or *prime*)
3. En garde

Oppose in *sixte* and riposte below with a *croisé*

Beat, disengage

Press, disengage

At the change of engagement, disengage in time

Change the line

Low line parries after stepping (or an attack on the step), without (or with) double engagement and simple ripostes.

On one or two disengagements:

---

[82]dérobement

1. Engage the sword or
   double the engagement):
   advance or retreat
2. Half-circle: oppose (or
   parry) and riposte (below
   or above
3. *En garde*

} Sometimes deceive
the engagement

Octave: oppose (or parry) and riposte (like in the first exercise).

Half-circle and *octave*: oppose (or parry) and riposte (like in the first exercise).

Half-circle and counter: oppose (or parry) and riposte (like in the first exercise).

Change the line.

Sixte and *octave*: oppose (or parry) and riposte (like in the first exercise).

Counter of *quarte* and half-circle: oppose (or parry) and riposte (like in the first exercise).

Change the line.

Third Iteration

Attacks deceiving the double engagement.

Straight strike.

1. Engage the sword (double the engagement without covering oneself)[83]
2. Deploy the arm
3. Lunge
4. *En garde*

} Or without deconstructing: strike straight

Disengage (on the double engagement covering oneself)
Double
Double, disengage
One, two
One, two, deceive
One, two, three
Change the line

Attacks on the absense of the sword in riposte and in attack

---

[83] *doubler l'engagement sans se couvrir*

Remise

1. Engage the sword
2. Disengage (make an absense of the sword in order to riposte)
3. *Remise*
4. *En garde*

Change the line

Reprise

1. Engage the sword
2. Disengage (parry with riposting)
3. Retake the straight strike (or the cutover)
4. *En garde*

Change the line.

Timed Strike.

1. Engage the sword
2. On the feint (or absence of the sword) strike straight.
3. *En garde.*

Change the line.

Sixth Lesson

First Iteration

Simple counter-ripostes after several parries and simple ripostes

1. Engage the sword.
2. Disengage (parry and riposte above or below).
3. *Sixte* (or *octave*): oppose (or parry) in lifting yourself[84]
4. Strike above (or below).
5. *En garde.*

Note: The counter-ripostes will be made above if the riposte has take place above and, reciprocally, have the student's parries varied on the disengagement, according to the following order.

The order of the parries taken by the instructor on the student's disengagement and ripostes.

---

[84] *opposez (ou parer) en vous relevant*

1. *Sixte*, above and below.
2. *Quarte*, above and below.
3. Counter of *sixte*, above and below.
4. Counter of *quarte*, above and below.
5. Half-circle, above and below.
6. *Octave*, above and below.
7. *Tierce*, above and below.
8. *Seconde*, above and below.
9. *Quinte*, above and below.
10. *Prime* below.
11. *Prime* and cutover, above.
12. *Prime*, cutover.

Second Iteration

Simple counter ripostes on the firm foot, after simple parries and compound ripostes.

Straight strike.

1. Engage the sword.
2. Feint a disengagement (parry or straight riposte): variable.
3. *Sixte* (or counter of *quarte*): oppose (or parry): counter: straight riposte: unchanging[85]
4. *En garde.*

Ripostes on the Feint

Disengage.
  Double.
  One, two.
  Cutover.
  Circling the sword.
  Change the line.

Compound counter ripostes, on the firm foot, after parries and simple ripostes.

Straight strike

1. Engage the sword.
2. Feint of disengagement (parry and straight riposte). Unchanging.

---

[85] *invariable*

3. *Sixte* (or counter of *quarte*): oppose (or parry): counter: straight riposte. Variable.
4. *En garde.*

Counter riposte after the parry

Disengage.
  Double.
  One, two.
Cutover.
Circling the sword. } Change the line.

Third Iteration

Parries from the counter riposte and second simple counter riposte
Parries and simple ripostes, compound counter ripostes

1. Engage the sword
2. *Sixte* (or counter of *quarte*): oppose (or parry). Straight riposte.
3. The counter riposte: straight strike (changeable or invariable)
4. *En garde.*

Disengage.
Double.
One, two.
Cutover.
Circling the sword.
Change the line.

Parries from the counter riposte and second compound counter ripostes

Parries, ripostes and simple counter ripostes

1. Engage the sword
2. *Sixte* (or counter of *quarte*): oppose (or parry). Straight riposte. Invariable.
3. *Sixte* (or counter of *quarte*): oppose (or parry). Straight strike. Changeable.
4. *En garde.*

Disengage.
Double.

One, two.
Cutover.
Circling the sword.
Change the line.

Seventh Lesson

First Iteration

An opposition. Straight strike and disengagement.
    Two oppositions.
    Three oppositions.
    Change the line.
    A counter (straight strike and disengagement).
    Two counters.
    Change the line.

Second Iteration

An opposition and a counter.
    An opposition, a counter and an opposition.
    An opposition and two counters.
    Two oppositions and a counter.
    A counter and an opposition.
    A counter and two oppositions.
    A counter, an opposition and a counter.
    Change the line.

Third Iteration

Repeat the first iteration above preceding each attack with a double disengagement.

Related Observations

1. **At the *En Garde*[86].** The instructor will endeavour to give the student perfect balance and, to this effect, he will ensure that the feet are not crossed, that the body is straight, plumb on both hips and without stiffness, that the shoulders are well cleared[87]

---

[86] *à la mise en garde*
[87] *les épaules soient bien effacées*

and a little open [and], finally, that the position of the arms and of the right hand is not disordered,[88] as happens always in beginners. He will often reposition the student and will endeavour to give him a guard position so natural and so easy that he could maintain it for a long time without fatigue.

2. **On the step**. To ensure that the men maintain good balance in advancing and retreating,[89] the instructure will command "Two *appels*".

3. **In deploying the arm**. This exercise has for its aim relaxing the arm[90] and avoiding the stiffness in the shoulder so common in beginners. In order to obtain this result, the instructor will ensure above all that the body does not follow the movement of the arm when it is extended or flexed.

4. **In the lunge**. The student being in the habit, in the beginning, of letting the body go[91] in lunging and lowering the head, the instructor must watch with attention the stiffness of the left knee,[92] the position of the body and that of the left shoulder, which must be well removed. He must equally observe the student not lunge too far, of maintaining the left foot flat. Without this, he can no longer raise himself quickly. In order to ensure that the student carries well the weight of the body on the left leg, the instructor will make him give two *appels* with the foot before lunging.

5. **In the attack**. Ensure the attack is always executed thoroughly and with all possible speed. Despite the fastest recorded, to which we should always strive, the arm must be deployed before the lifting of the foot and the wrist must be carried to the side of the opposing sword in order to cover oneself only in the moment of the final lunge of the strike hitting, the only moment where the riposte is to be feared.

6. **In the Counters**. The lesson of the counters is the application of the lesson to the assault.[93] It has as its aim allowing the acquisition of the speed necessary for the execution with as greater

---

[88] *dérangée*

[89] *en marchant et en rompant*

[90] *ayant pout but d'assouplir le bras*

[91] *abandonner le corps*

[92] *la tension du jarret gauche*

[93] ie: the bout or sparring

ease as determined by the strike and the parry to be executed removing the usual hesitation occasioned by the unexpected in the bout. It is performed in pairs:[94] one of the two guides it with the attack and, in this case, changes the exercises when he judges it appropriate, or deceiving the parry, or changing his last.

Article VI — The Wall and Sparring[95]

Wall: The wall is the prelude to sparring. It consists of some conventional disengagements and parries executed with the greatest smoothness[96] in order to prepare the hand and legs. It is accompanied by a salute to the gallery and to the opponent.

The wall and the salutes are excecuted in pairs in the following manner:

The Wall (in six tempos)

First tempo

1. Lower the right hand from *tierce* and raise the left arm.
2. Flex (as in taking guard)
3. Advance the foot (as in taking guard)
4. Gather backwards

Note: Without detailling, invite the attack by these words: "The honour goes to you"[97] or respond "In obedience,"[98] and fence.

Second tempo

1. Lunge
2. Raise yourself

---

[94] *elle est exécutée à deux*
[95] *Mur et Assaut*
[96] *la plus grande régularité*
[97] *À vous l'honneur!*
[98] *Par obéssiance*

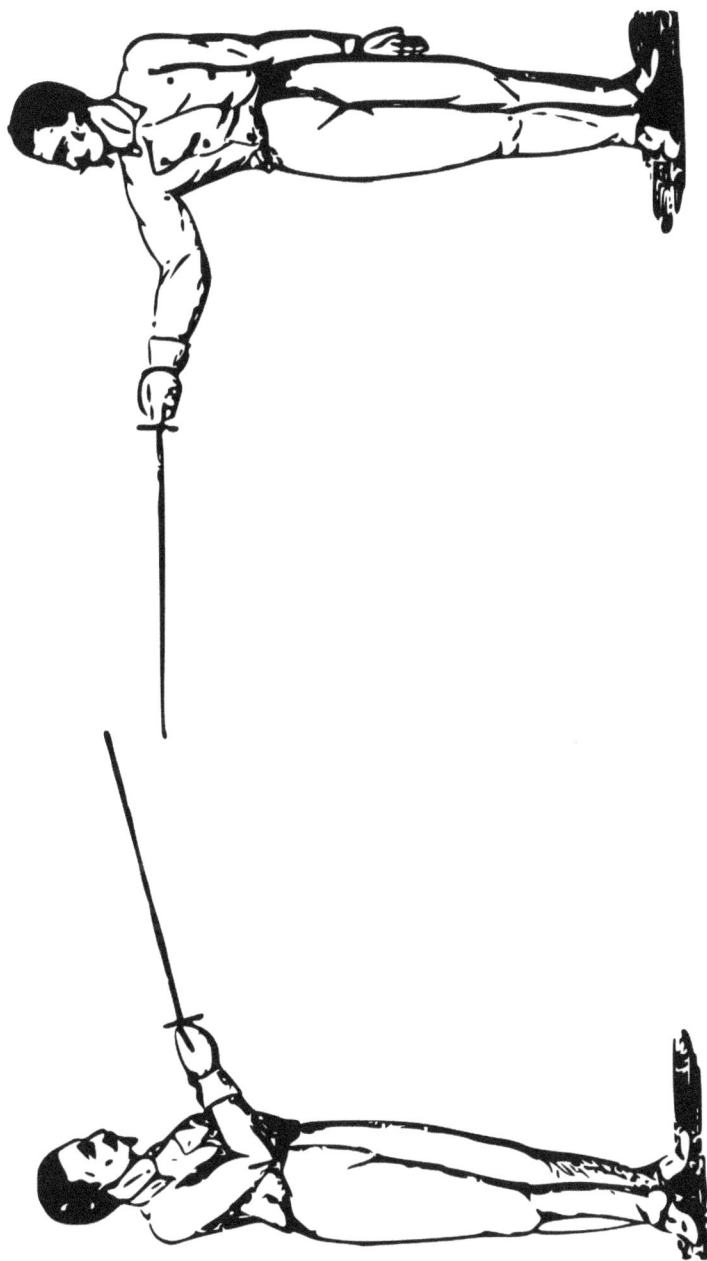

Fig. 21 and Fig. 22

3. Salute
{
1. Salute to the left, arm half extended, the fingers upwards, the head turned to the left (fig 21).
2. Salute to the right, the arm half extended, the fingers downwards, the head turned to the right (fig 22).
3. Return to the first position.
}

4. *En garde* (Put yourself in guard in seven tempos).

Third tempo

1. Disengage (the opponent parries with *tierce*)(fig. 23).
2. *En garde* (retake the guard and touch swords).
3. Disengage (the opponent parries in *quarte*)(fig. 24).
4. *En garde* (retake the guard and touch swords).
5. One, two, gather forward (the opponent parries in *tierce*).
6. Lower the right hand from *tierce* and raise the left arm
7. Escape[99] backwards (putting oneself into guard backwards)
8. Two *appels*
9. Gather backwards.

Fourth tempo

As the second tempo.

Fifth tempo

As the third tempo.

Sixth tempo

1. Lower the right hand from *tierce* and raise the left arm
2. Escape backwards
3. Two *appels*
4. Salute (as in the third tempo gathering forward)

5. Salute in front of you
{
1. Curve the arm, the elbow joined to the body, the hand at the height of the chin, the fingers turned towards the body (fig.25).
2. Lower the blade extending the arm, the fingers of the right hand upwards and to the side of the right thigh (fig 26).
}

---

[99] *escapper*

Fig. 23

Fig. 24

Fig. 25 and Fig. 26

Sparring[100]

Sparring is the strict and reasoned application of the rules of fencing by a pair of simulated adversaries, armed with masks and gloves: it is the appearance of combat. It teaches, with timeliness, the best parts of the lessons taught at the *plastron*.

The student must only be allowed to spar[101] when he is suitably affirmed in the principles: when he executes well the counters, when the movements of all part of his body are well regulated and when his hand and legs have acquired the speed, precision and the togetherness indespensible to good performance.[102]

Sparring does not preclude lessons. On the contrary, one must continue at the *plastron*, not only to progress but also in order to maintain and preseve that which one has learned.

Advice and Observations

1. Put yourself into guard out of distance[103] in order to avoid any surprises; but always crossing swords.

2. Advance in little steps and always ready to parry in order to put oneself into distance[104] and attack more easily.

3. Retreat to maintain even distance or in order to step and attack on the step.

4. Attack rather with simple direct strikes or preceed with a beat or pressing, and, as much as possible, on preparations such as: engagements, step, retake guard after the attack, absences of the sword whether in attack or whether in parrying. Attack with compound strikes only when the simple strikes or well executed feints have been parried and, by this fact, will destroy any thought of a stop-strike.[105]

5. Raise yourself quickly, whether or not the strike hit, in order to more easily take the defensive and above all in order to escape grappling.[106]

---

[100] *Assaut*
[101] *à faire assaut*
[102] *l'ensemble indispensables à toute bonne exécution*
[103] *hors de portée*
[104] *se mettre à portée*
[105] *coup d'arret*
[106] *le corps à corps*

6. Parry in variety often.[107] Only employ the simple and most useful but make the hand easy to move from side to side.[108] The counter gives more stability and returns if one is extended.

7. Riposte always after the single parry[109] and not in executing it. The simple riposte is the best. It happens quickest and prevents the *remise* and *reprise*.

8. Seek to see cleanly, to determine[110] wisely and to act quickly in order to acquire timeliness and directness[111] in the attack, precision and surety in the parry and calculated calmness[112] in the riposte, of which relative speed must be the inference of rapid reasoning.

9. In order to fence with all his means, the bout must not exceed fifteen minutes.

10. Always spar[113] in the presence of a master or provost, who, after the beautiful and final salute, will analyse the incidents in the fight, remark on the faults committed, indicate how one could avoid them and make each opponent understand the specific defects[114] in his game and the means to remedy them in order to acquire the judgement and timeliness in the strike to hit[115] and parry.

Rules to observer during the bout:

1. Do not strike without having crossed swords.

2. Never push on the strike when one has touched and never accentuate it by cries in bad taste.

3. Do not riposte if the parry has diarmed the opponent, except in the case of the riposte from a *tac-au-tac*, that is to say, from the straight strike.

---

[107] *parer en variant souvent*
[108] *ébranler* in the sense of oscillation, eg: *ébranler une cloche*
[109] *la parade seulement*
[110] *à deviner*
[111] *franchise*
[112] *le sang-froid calculé*
[113] *exécuter l'assaut*
[114] *les défectuosités caractéristiques*
[115] *à porter*

4. Never ask if the strike succeeded in touching. Each time that one is hit, declare it faithfully saying "*touché*".

5. Do not argue while fencing. Leave the evaluation and judgement of the strikes to the gallery.

6. Do not complain about the opponent's play. If he tires or is not up to it,[116] make an honest excuse to end the bout.

---

[116] *ne convient pas*

Second Part: Fencing with the Sabre or *Contre-Point*

Article I — Outline the method of teaching. The spirit in which it must be practiced.

The principles and the spirit of the method of teaching are the same for fencing with the sabre as for fencing with the *épée*. Refer, therefore, to the first article in the first part.

Article II — Description of the Sabre and the positions and the movements which are attached to putting oneself into play for the attack and the defence.

Nomenclature of the sabre

The *sabre*, tool of the *contre-point*, is composed of two principle parts, the *blade* and the *mounting*.

The blade presents: the *point*, the *back*, the *cutting edge*, the *heel* and the *tang* engaged in the mounting. The mounting comprises the *handle* and the *guard*.

Fig. 1

The Manner of Holding the Sabre

The handle in the right hand, the fingers below, the thumb along the back of the sabre and nearly touching the guard, the four other fingers together underneath and gripping the handle lightly, the edge to the right, the heel to the outside and on the side of the handle.

Preparatory Movements

Being in the position of the unarmed soldier, make a half-turn maintaining the head straight, the feet placed squarely, without separating the heels, the right arm outspread before and separated from the body, the point of the sabre about 0.10m from the ground, the left forearm joined to the body and deployed behind (fig. 1).

Fig. 2

Guard

1. Raise the sabre, the arm extended, the hand to the height of the eyes (fig. 2).

2. Bend the right arm such that the hand may be at the height of the right breast, the wrist opposite the right shoulder, the fingers downwards, the elbow to the outside to the right and a little detached from the body, the point of the sabre at the height of the eyes (fig. 3).

3. Bend the legs and advance the right foot (fig. 4) as with fencing with the *épée*. Make the second movements of the two moulinets, the first on the left as soon as these have been demonstrated.

Fig. 3

Fig. 4

Fig. 5

Step, *Appels*, Gathering, Lunge

Execute after the principles in the teaching of the *épée*.

The lunge is always preceded by the deployment of the arm made with the aid of a moulinet.

Voiding the Leg

The cuts made to the leg or the thigh are not parried with the sabre but by slipping, that is to say, quickly carrying the right leg stretched rearwards around 0.35m, the foot flat (fig. 5).

Exercises for Limbering Up the Arm and the Wrist

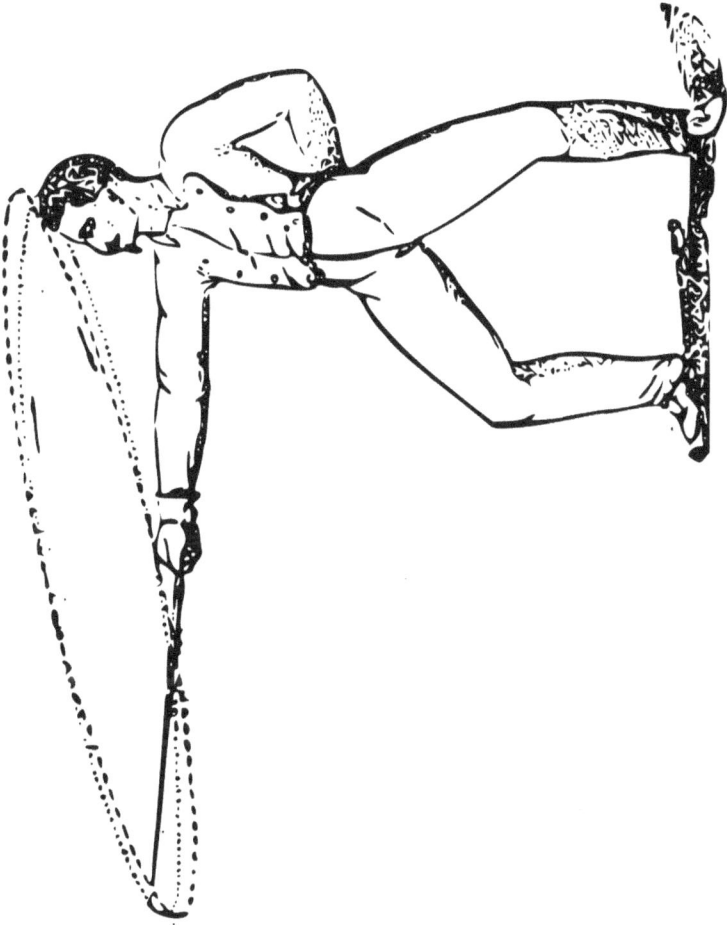

Fig. 6

Moulinets

Extend the right arm forward, the hand at the height of the shoulder, fingers downwards. Make the sabre describe a circle, horizontally, above the head, from left to right (or from right to left) opening lightly the fingers, the wrist, to help the movement, and replace the hand, fingers upwards (or downwards), the edge of the sabre to the left (or to the right) (fig. 6).

Engagement and Lines

Fig. 7

Cross swords, edge against edge, carrying the wrist to the right (or to the left), the fingers upwards (or downwards) in order to be covered.

Of the two engagements, that to the right (fig. 7), with the position of the hand facilitating the attack and above all the defence of the low line, has greater application.

The strikes and parries are demonstrated in this instruction, starting from the engagement on the right.

Refer, for the definition of the lines, to the teaching of fencing with the *épée*.

Attacks

The attack may be made by a simple strike or by a compound strike not exceding three movements. The simple strikes are done:

By a moulinet for head strikes and sash strikes;[117]

By a moulinet for face strikes to the right and to the left;

[117] *pour les coups de tête et de banderole*

By a sabre strike[118] for the flank, stomach and forearm strikes and for the point.

The feint serves to form compound strikes. It is executed as the strike itself but without lunging.

Head Strike

Execute a moulinet backwards to the left and deploy the arm stopping the sabre at the height of the top of the head, the edge forward (fig. 8).

Sash Strike[119]

Execute a moulinet backwards to the left and deploy the arm stopping the sabre at the height of the shoulder, the edge forward, directing the strike diagonally from right to left (fig. 9).

Face Strike to the Right

Execute a moulinet from right to left and deploy the arm stopping the sabre at the height of the face, the edge to the right, the hand fingers downwards (fig. 10).

Face Strike to the Left

Execute a moulinet from left to right and deploy the arm, stopping the sabre at the hieght of the face, the edge to the left, the hand fingers upwards (fig. 11).

Flank Strike

Deploy the arm stopping the sabre at the height of the flank, the edge upwards, the thumb slightly to the left (fig. 12).

Stomach Strike

Deploy the arm stopping the sabre at the height of the stomach, the edge upwards, the thumb slightly to the right (fig.13).

Fig. 8

Fig. 9

Fig. 10

Fig. 11

Fig. 12

Fig. 13

Forearm Strike

Make the sabre strike, the edge downwards, the thumb slightly to the right so as to avoid, by stopping the forearm, the execution of the head strike.

Point Strike

Lower the point of the sabre to the height of the chest and deploy the arm turning the hand, the thumb downwards, the edge of the sabre upwards (fig. 14).

The Parry

The parry is always made by the opposition of the edge against edge and leaves the sabre on the side where it is presented, chasing it from the body without returning it to the point of departure. It is divided into:

1. The head parry, encompassing the head and face;

2. The body parry, encompassing the sash, flank, stomach and point.

Head Parry

Raise the right arm turning the hand, the fingers forward, and place the sabre horizontally a little before and at the height of the top of the head, the edge upwards

Face Parry to the Right (or to the Left)

Carry the wrist to the right (or to the left) opposite and about 0.10m from the breast, the blade of the sabre inclined slightly forward, the edge to the right (or to the left).

Sash and Stomach Parry

Raise the right arm, the elbow to the outside, bending it, and place the forearm horizontally before the body and at the height of the shoulder, the hand, fingers in front, and opposite the middle of the body,

---

[118] *la coup de sabre*
[119] *Coup de banderole*

Fig. 14

the point of the sabre low, the blade about 0.10m from the body, the edge to the left.

Flank Parry

Carry the wrist outside and to the right, bending lightly the arm, the elbow and the hand at the height of the shoulder, the point of the sabre low, the blade around 0.33m from the body, the edge to the right.

Point Parry

Being in guard, incline slightly the point, the wrist opposite the middle of the body.

Riposte

The riposte can be made with the same strike as the attack.

It must be directed to the uncovered side of the body, thus guaranteeing the most likely[120] side attacked. It will be executed in the following manner:

Attack – Riposte

Attack: to the head
Riposte: to the stomach or to the flank

Attack: to the face to the right
Riposte: to the face to the left or to the flank

Attack: to the face to the left
Riposte: to the face to the right or to the stomach

Attack: to the sash or to the stomach
Riposte: to the head or to the sash

Attack: to the flank
Riposte: to the head or to the face to the right or to the stomach

Attack: from the point
Riposte: to the head or to the face to the right

These same strikes can be made in counter-riposte.

The riposte is executed also in turning the wrist in place after the parry, without leaving the sabre opposed, in order to put anew the edge forward and double the sabre strike in the following manner:

---

[120] *le plus possible*

Attack – Riposte
> Attack: to the face to the left
> Riposte: to the face to the right and to the sash

> Attack: to the face to the right
> Riposte: to the face to the left and to the flank

> Attack: to the flank
> Riposte: to the stomach and to the face to the right

> Attack: to the sash
> Riposte: to the flank and to the face to the left

> These strikes can be made in counter-riposte.

Article III — Teaching progression

Teaching is given in four lessons of three iterations each.

First Lesson

First iteration: Preparatory movements and taking guard
> Second iteration: Moulinets on the firm foot
> Third iteration: Moulinets lunging

Second Lesson

First iteration: Attacks by simple strikes
> Second iteration: Parries after a simple attack
> Third iteration: Meeting the simple attack with the parry

Third Lesson

First iteration: Attacks by compound strikes
> Second iteration: Parries after a simple attack and simple ripostes
> Third iteration: Parries after a simple attack and compound ripostes

Fourth Lesson

First iteration: Varied attacks[121]

---

[121] *Attaques diverses*

Second iteration: Parries after a compound attack and simple ripostes

Third iteration: Parries after a compound attack and compound ripostes

Article IV — General Rules to Observe

The general rules to observe are the same as for fencing with the *épée*. Refer to Article IV in the first part.

Article V — Lesson Detail

First Lesson

First Iteration

Preparatory movements

1. $\begin{cases} \text{1. Take the guard in four steps} \\ \text{2. } \textit{En garde} \end{cases}$
2. Gather forward (or backwards)
3. Step forward (or retreat)

Note: Execute the *appels* and the salute.

Second iteration

Moulinets on the firm foot

1. $\begin{cases} \text{1. For the moulinets to the right (or to the left). In position.} \\ \text{2. Begin} \\ \text{3. Stop} \end{cases}$
2. Moulinets to the left (or to the right)
3. Moulinets to the left or the right (or to the right and to the left)

Third iteration

Moulinets lunging

1. $\begin{cases} \text{1. With a moulinet to the right (or to the left)} \\ \quad \text{Lunge} \\ \text{2. } \textit{En garde} \end{cases}$

2. $\begin{cases} \text{With a moulinet to the right (or to the left)} \\ \text{Slip [the leg]} \end{cases}$

Second Lesson

First iteration

Attacks with simple strikes

1. $\begin{cases} \text{1. For the head strike: Lunge} \\ \text{2. } En\ garde \end{cases}$
2. For the left face strike
3. For the right face strike
4. For the flank strike
5. For the stomach strike
6. For the sash strike
7. For the point strike
8. $\begin{cases} \text{1. For the forearm strike} \\ \text{2. Slip [the leg]} \end{cases}$

Second iteration

Parries of simple strikes

1. $\begin{cases} \text{1. The head strike: Parry} \\ \text{2. } En\ garde \end{cases}$
2. For the left face strike
3. For the right face strike
4. For the flank strike
5. For the stomach strike
6. For the sash strike
7. For the point strike

Third iteration

Meeting the simple attack with the parry
Note: the parry must be made rising.[122]

---

[122] *La parade doit être faite en se relevant*

1.
1. With a sash strike: Lunge
2. For the head: Parry
3. For the point strike: Lunge
4. *En garde*

2.
1. With a flank strike: Lunge
2. For the head: Parry
3. With a stomach strike: Lunge
4. For the head: Parry
5. With a point strike: Lunge
6. *En garde*

3.
1. With a head strike: Lunge
2. For the stomach: Parry
3. With a sash strike: Lunge
4. For the head: Parry
5. With a point strike: Lunge
6. *En garde*

4.
1. With a left face strike: Lunge
2. For the right face strike: Parry
3. With a stomach strike: Lunge
4. For the sash: Parry
5. With a sash strike: Lunge
6. For the head: Parry
7. With a point strike: Lunge
8. *En garde*

5.
1. With a right face strike: Lunge
2. For the flank strike: Parry
3. With a head strike: Lunge
4. For the stomach: Parry
5. With a sash strike: Lunge
6. For the head: Parry
7. With a point strike: Lunge
8. *En garde*

6. {
1. On the head strike, with a forearm strike: Slip
2. With the stomach strike: Lunge
3. For the sash: Parry
4. With a sash strike: Lunge
5. For the head strike: Parry
6. With the flank strike: Lunge
7. For the head: Parry
8. With a point strike: Lunge
9. *En garde*

Third Lesson

First iteration

Attacks with compound strikes

1. {
1. Point strike
2. Lunge
3. *En garde*

Note: Repeat the same exercise several times in succession, allowing to touch each time. To end, parry and throw a point strike. This rule is common to all exercises.

2. {
1. Feint a point strike
2. Head strike

3. {
1. Feint with the point
2. Right face strike

4. {
1. Feint with the point
2. Left face strike

5. {
1. Feint with the point and with the left face strike
2. Stomach strike

6. {
1. Feint with the point and with the left face strike
2. Right face strike

7. {
1. Feint with the point and with the right face strike
2. Left face strike

8. {
1. Feint with the point and with the left face strike
2. Flank strike

9. {
1. Feint with the point and with the right face strike
2. Flank strike

10. $\begin{cases} \text{1. Feint with the point and with the head [strike]} \\ \text{2. Flank strike} \end{cases}$

11. $\begin{cases} \text{1. Feint with the point and with the head [strike]} \\ \text{2. Sash strike} \end{cases}$

Second iteration

Simple parries after a simple strike and simple ripostes

1. $\begin{cases} \text{1. Parry the point strike} \\ \text{2. With a right face strike: Riposte} \\ \text{3. } \textit{En garde} \end{cases}$

2. $\begin{cases} \text{1. Parry the head strike} \\ \text{2. With a stomach strike: Riposte} \end{cases}$

3. $\begin{cases} \text{1. Parry the right face strike} \\ \text{2. With a left face strike: Riposte} \end{cases}$

4. $\begin{cases} \text{1. Parry the left face strike} \\ \text{2. With a right face strike: Riposte} \end{cases}$

5. $\begin{cases} \text{1. Parry the stomach strike} \\ \text{2. With a head strike: Riposte} \end{cases}$

6. $\begin{cases} \text{1. Parry the flank strike} \\ \text{2. With a right face strike: Riposte} \end{cases}$

[7.] $\begin{cases} \text{1. Parry the sash strike} \\ \text{2. With a head (or sash) strike: Riposte} \end{cases}$

Third iteration

Simple parries after a simple strike and compund ripostes

1. $\begin{cases} \text{1. Parry the left face strike} \\ \text{2. With right face and stomach strikes: Riposte} \end{cases}$

2. $\begin{cases} \text{1. Parry the flank strike} \\ \text{2. With stomach and right face strikes: Riposte} \end{cases}$

3. $\begin{cases} \text{1. Parry the right face strike} \\ \text{2. With right face and flank strikes: Riposte} \end{cases}$

4. $\begin{cases} \text{1. Parry the stomach strike} \\ \text{2. With flank and left face strikes: Riposte} \end{cases}$

Fourth Lesson

First iteration

Varied attacks

1. {
   1. Feint a point strike above
   2. Stomach strike

2. {
   1. Feint a right face strike (above the point)
   2. Stomach strike

3. {
   1. In changing engagement: Beat (turning the hand)
   2. Right face strike (replacing the hand)

4. {
   1. On the head strike: Forearm strike with a moulinet
   2. Forearm strike with a moulinet

5. {
   1. Parry the point strike
   2. Onto my head strike, point strike

Second iteration

Simple ripostes and parries after compound strikes

1. {
   1. Parry the head and point strikes
   2. With a right face strike: Riposte

2. {
   1. Parry the point and head strikes
   2. With a stomach strike: Riposte

3. {
   1. Parry the point and right face strikes
   2. With a left face strike: Riposte

4. {
   1. Parry the point and left face strikes
   2. With a right face strike: Riposte

5. {
   1. Parry the point high and stomach strikes
   2. With a head strike: Riposte

6. {
   1. Parry the point, face and flank strikes to the left or to the right
   2. With a right face strike: Riposte

Third iteration

Compound parries and ripostes after simple strikes

1. {
  1. Parry the point and right face strikes
  2. With left face and flank strikes: Riposte

2. {
  1. Parry the point and left face strikes
  2. With right face and stomach strikes: Riposte

3. {
  1. Parry the point, left (or right) face and flank strikes
  2. With stomach and flank strikes: Riposte

4. {
  1. Parry the point, head and sash strikes
  2. With flank and left face strikes: Riposte

Article VI — Salute and the Bout

Salute

The salute preceeding the bout is executed in the following manner:

1. Execute a lunging right face strike.[123]
2. Come back to the position of the first movement of the guard.
3. Put yourself *en garde*.
4. Gather forward crossing sabres.
5. Execute two changes of guard.
6. Slip backwards.
7. Two *appels*.
8. Salute to the right and to the left, and gather forward.
9. Put yourself *en garde*.
10. Invite the attack with these words: "The honour [goes] to you"[124] and let yourself be touched, or respond "in obedience."[125] Lunge with a flank (or point strike) and raise yourself back into guard.
11. Salute to the right, gathering backwards.
12. Salute in front of you.

The Bout

In application, endeavour to give the strikes of the sabre by movements close to the body, in a manner to disturb as little as possible the wrist off the line and to make the strikes with lightness.

---

[123] *un coup de figure à droite en se fendant*

[124] *à vous l'honneur*

[125] *par obeissance*

After having touched, retire the sabre quickly backwards, impressing on it an oblique direction, in the direction of the edge, in the manner of a saw.

Rules to Observe during the Bout

Refer to Article VI of the first part.

Conduct and Discipline in the *Salle d'Armes*. Duties of the Masters and Provosts.

The masters and provosts must have to heart constantly holding the *salle d'armes* in the best state of cleanliness and looking to decorate it with the help of panoplies [of weapons] and sheaves of foils,[126] masks and gloves, topped with panels in cardboard[127] or wood recalling the battles and combats in which the regiment, the battalion or squadron has taken a glorious part, or by highlighting noble devices such as "Honour and Homeland. Valour and Discipline. Courage and Dedication. Glory to God. Respect to the Masters. Honour in Arms. *Vive la France*", etc.

They must endeavour to have always observed in the *salle* the courtesy, decency and marks of respect prescribed by the Internal Service[128] and prevent smoking there and require that one enters there uncovered.

They must take care to prevent quarrels or to appease them and to only intervene in the aim of conciliation in the name of the military confraternity.

They will never tolerate anyone to spar[129] outside of their presence and without their authorisation which they must accord only with prudence and discernment, and only in the interests of instruction.

They must follow with attention all the events of the bout which they have authorised, with the method of being able, after the clear[130] and final salute, to remark on the faults committed, indicate how one could avoid them and signal to each opponent the characteristic defects in his game and the means to remedy them.

---

[126] *faisceaux de fleurets*
[127] pasteboard?
[128] *le service intérieur*
[129] *qu'on fasse assaut*
[130] *belle?*

They must, in all circumstances, give the example of calm and measured language, of a posture correct and sparkling,[131] of a faithful and irreproachable conduct and of maintaining dignity and pride, imposing respect to the uniform and inspiring esteem and consideration for their own person.

---

[131] *d'une tenue correcte et brillante*

Also from LongEdge Press

LongEdge Press publishes quality translations of French texts of interest to the scholar and practitioner of historical fencing. Visit LongEdge Press at www.longedgepress.com.

Secrets of the Sword Alone Henri de Sainct-Didier (1573)

Fencing Through the Ages Adolphe Corthey (1898), including his report on the *Transformation de l'Épée de Combat* (1894) and several contemporary newspaper reports of his public demonstrations of historical fencing.

La Canne Royale Larribeau, Humé, comprising Larribeau's *Nouvelle Théorie du Jeu de la Canne* (1856) and Humé's *Traité et Théorie de Canne Royale* (1862)

The Art of Fencing Jean de Brye (1721)

Manual of *Contre-Pointe* Fencing Joseph Tinguely (1856) with an foreword by Julian Garry of *De Taille et d'Estoc*